TYCHO BRAHE

GREAT ASTRONOMERS

ROBERT STAWELL BALL

GIFT *Certificate*

TO:

____

____

FROM:

____

____

DATE: ______________________

TYCHO BRAHE

GREAT ASTRONOMERS

BY

ROBERT STAWELL BALL

D.Sc, LL.D., F.R.S.

(July 1, 1840 – November 25, 1913)

LOWNDEAN PROFESSOR OF ASTRONOMY AND GEOMETRY IN THE UNIVERSITY OF CAMBRIDGE

DIAMOND BOOKS

www.diamondbooks.ca

TORONTO, CANADA – 2017

DIAMOND BOOKS - CANADA

DIAMOND BOOKS

Toronto, ON, CANADA

http://www.diamondbooks.ca

BIBLIOGRAPHIC INFORMATION

'TYCHO BRAHE'
Is an extract or a chapter from the book
"GREAT ASTRONOMERS" by Robert Stawell Ball.
It was first published in 1896.

PUBLISHED IN CANADA

Published in Canada by DIAMOND BOOKS - CANADA, an imprint of DIAMOND PUBLISHERS - http://www.diamondpublishers.com

First Edition: March, 2017.

PAPERBACK EDITION : ISBN: 978-1-988357-52-2
E-BOOK EDITION : ISBN: 978-1-988357-53-9

PRINTED IN CANADA

PREFACE
TO THE ORIGINAL EDITION

It has been my object in these pages to present the life of each astronomer in such detail as to enable the reader to realise in some degree the man's character and surroundings; and I have endeavoured to indicate as clearly as circumstances would permit the main features of the discoveries by which he has become known.

There are many types of astronomers—from the stargazer who merely watches the heavens, to the abstract mathematician who merely works at his desk; it has, consequently, been necessary in the case of some lives to adopt a very different treatment from that which seemed suitable for others.

ROBERT STAWELL BALL.

The Observatory, Cambridge.
October, 1895.

NOTE: The full-length of 'Preface' can be found in its original edition of the book "GREAT ASTRONOMERS".

INTRODUCTION

Of all the natural sciences there is not one which offers such sublime objects to the attention of the inquirer as does the science of astronomy. From the earliest ages the study of the stars has exercised the same fascination as it possesses at the present day. Among the most primitive peoples, the movements of the sun, the moon, and the stars commanded attention from their supposed influence on human affairs. The practical utilities of astronomy were also obvious in primeval times. Maxims of extreme antiquity show how the avocations of the husbandman are to be guided by the movements of the heavenly bodies. The positions of the stars indicated the time to plough, and the time to sow. To the mariner who was seeking a way across the trackless ocean, the heavenly bodies .offered the only reliable marks by which his path could be guided. There was, accordingly, a stimulus both from intellectual curiosity and from practical necessity to follow the movements of the stars. Thus began a search for the causes of the ever-varying phenomena which the heavens display.

\- ROBERT STAWELL BALL

Many of the earliest discoveries are indeed prehistoric. The great diurnal movement of the heavens, and the annual revolution of the sun, seem to have been known in times far more ancient than those to which any human monuments can be referred. The acuteness of the early observers enabled them to single out the more important of the wanderers which we now call planets. They saw that the star-like objects, Jupiter, Saturn, and Mars, with the more conspicuous Venus, constituted a class of bodies wholly distinct from the fixed stars among which their movements lay, and to which they bear such a superficial resemblance. But the penetration of the early astronomers went even further, for they recognized that Mercury also belongs to the same group, though this particular object is seen so rarely. It would seem that eclipses and other phenomena were observed at Babylon from a very remote period, while the most ancient records of celestial observations that we possess are to be found in the Chinese annals. The study of astronomy, in the sense in which we understand the word, may be said to have commenced under the reign of the Ptolemies at Alexandria. The most famous name in the science of this period is that of Hipparchus, who lived and worked at Rhodes about the year 160 B.C. It was his splendid investigations that first wrought the observed facts into a coherent branch of knowledge. He recognized the primary obligation which lies on the student of the heavens to compile as complete an inventory as possible of the objects which are there to be found. Hipparchus accordingly commenced by undertaking, on a small scale, a task exactly similar to that on which modern astronomers, with all available appliances of meridian circles, and photographic telescopes, are constantly engaged at the present day. He compiled a catalogue of the principal fixed stars, which is of special value to astronomers, as being the earliest work of its kind which has been handed down. He also studied the movements of the sun and the moon, and framed theories to account for the incessant changes which he saw in progress. He found a much more difficult problem in his attempt to interpret satisfactorily the complicated movements of the planets. With the view of constructing a theory which should give some coherent account of the subject, he made many observations of the places of these wandering stars. How great were the advances which Hipparchus accomplished may be appreciated if we reflect that, as a preliminary task to his more purely astronomical labours, he had to invent that branch of mathematical science by which alone the problems he proposed could be solved. It was for this purpose that he devised the indispensable method of calculation which we now know so well as trigonometry. Without the aid rendered by this beautiful art it would have been impossible for any n ally important advance in astronomical calculation to have been effected.

But the discovery which shows, beyond all others, that Hipparchus possessed one of the master-minds of all time was the detection of that remarkable celestial movement known as the precession of the equinoxes. The inquiry which conducted to this discovery involved a most profound investigation, especially when it is remembered that in the days of Hipparchus the means of observation of the heavenly bodies were only of the rudest description, and the available observations of earlier dates were extremely scanty. We can but look with astonishment on the genius of the man who, in spite of such difficulties, was able to detect such a phenomenon as the precession, and to exhibit its actual magnitude. I shall endeavour to explain the nature of this singular celestial movement, for it may be said to offer the first instance in the history of science in which we find that combination of accurate observation with skilful interpretation, of which, in the subsequent development of astronomy, we have so many splendid examples.

The word equinox implies the condition that the night is equal to the day. To a resident on the equator the night is no doubt equal to the day at all times in the year, but to one who lives on any other part of the earth, in either hemisphere, the night and the day are not generally equal. There is, however, one occasion in spring, and another in autumn, on which the day and the night are each twelve hours at all places on the earth. When the night and day are equal in spring, the point which the sun occupies on. the heavens is termed the vernal equinox. There is similarly another point in which the sun is situated at the time of the autumnal equinox. In any investigation of the celestial movements the positions of these two equinoxes on the heavens are of primary importance, and Hipparchus, with the instinct of genius, perceived their significance, and commenced to study them. It will be understood that we can always define the position of a point on the sky with reference to the surrounding stars. No doubt we do not see the stars near the sun when the sun is shining, but they are there nevertheless. The ingenuity of Hipparchus enabled him to determine the positions of each of the two equinoxes relatively to the stars which lie in its immediate vicinity. After examination of the celestial places of these points at different periods, he was led to the conclusion that each equinox was moving relatively to the stars, though that movement was so slow that twenty five thousand years would necessarily elapse before a complete circuit of the heavens was accomplished. Hipparchus traced out this phenomenon, and established it on an impregnable basis, so that all astronomers have ever since recognised the precession of the equinoxes) as one of the fundamental facts of astronomy. Not until nearly two thousand years after Hipparchus had made this splendid discovery was the explanation of its cause given by Newton.

From the days of Hipparchus down to the present hour the science of astronomy has steadily grown. One great observer after another has appeared from time to time, to reveal some new phenomenon with regard to the celestial bodies or their movements, while from time to time one commanding intellect after another has arisen to explain the true import of the facts of observations. The history of astronomy thus becomes inseparable from the history of the great men to whose labours its development is due. In the ensuing chapters we have endeavoured to sketch the lives and the work of the great philosophers, by whose labours the science of astronomy has been created.

We shall commence with Ptolemy, who, after the foundations of the science had been laid by Hipparchus, gave to astronomy the form in which it was taught throughout the Middle Ages. We shall next see the mighty revolution in our conceptions of the universe which are associated with the name of Copernicus. We then pass to those periods illumined by the genius of Galileo and Newton, and afterwards we shall trace the careers of other more recent discoverers, by whose industry and genius the boundaries of human knowledge have been so greatly extended. Our history will be brought down late enough to include some of the illustrious astronomers who laboured in the generation which has just passed away.

- ROBERT STAWELL BALL

TABLE OF CONTENTS

CHAPTERS

**An original Chapter from the book: "GREAT ASTRONOMERS", by Robert Stawell Ball.*

CHAPTER 3

GREAT ASTRONOMERS

Tycho Brahe

(December 14, 1546 – October 24, 1601)

Tycho Brahe

(December 14, 1546 – October 24, 1601)

CHAPTER 3

GREAT ASTRONOMERS: TYCHO BRAHE

THE most picturesque figure in the history of astronomy is undoubtedly that of the famous old Danish astronomer whose name stands at the head of this chapter. Tycho Brahe was alike notable for his astronomical genius and for the extraordinary vehemence of a character which was by no means perfect. His romantic career as a philosopher, and his taste for splendour as a Danish noble, his ardent friendships and his furious quarrels, make him an ideal subject for a biographer, while the magnificent astronomical work which he accomplished has given him imperishable fame.

- ROBERT STAWELL BALL

The history of Tycho Brahe has been admirably told by Dr. Dreyer, the accomplished astronomer who now directs the observatory at Armagh, though himself a countryman of Tycho. Every student of the career of the great Dane must necessarily look on Dr. Dreyer's work as the chief authority on the subject. Tycho sprang from an illustrious stock. His family had flourished for centuries, both in Sweden and in Denmark, where his descendants are to be met with at the present day. The astronomer's father was a privy councillor, and having filled important positions in the Danish government, he was ultimately promoted to be governor of Helsingborg Castle, where he spent the last years of his life. His illustrious son Tycho was born in 1546, and was the second child and eldest boy in a family of ten.

It appears that Otto, the father of Tycho, had a brother named George, who was childless. George, however, desired to adopt a boy on whom he could lavish his affection and to whom he could bequeath his wealth. A somewhat singular arrangement was accordingly entered into by the brothers at the time when Otto was married. It was agreed that the first son who might be born to Otto should be forthwith handed over by the parents to George to be reared and adopted by him. In due time little Tycho appeared, and was immediately claimed by George in pursuance of the compact. But it was not unnatural that the parental instinct, which had been dormant when the agreement was made, should here interpose. Tycho's father and mother receded from the bargain, and refused to part with their son. George thought he was badly treated. However, he took no violent steps until a year later, when a brother was born to Tycho. The uncle then felt no scruple in asserting what he believed to be his rights by the simple process of stealing the first born nephew, which the original bargain had promised him. After a little time it would seem that the parents acquiesced in the loss, and thus it was in Uncle George's home that the future astronomer passed his childhood.

When we read that Tycho was no more than thirteen years old at the time he entered the University of Copenhagen, it might be at first supposed that even in his boyish years he must have exhibited some of those remarkable talents with which he was afterwards to astonish the world. Such an inference should not, however, be drawn. The fact is that in those days it was customary for students to enter the universities at a much earlier age than is now the case. Not, indeed, that the boys of thirteen knew more then than the boys of thirteen know now. But the education imparted in the universities at that time was of a much more rudimentary kind than that which we understand by university education at present. In illustration of this Dr. Dreyer tells us how, in the University of Wittenberg, one of the professors, in his opening address, was accustomed to point out that even the processes of multiplication and division in arithmetic might be learned by any student who possessed the necessary diligence.

It was the wish and the intention of his uncle that Tycho's education should be specially directed to those branches of rhetoric and philosophy which were then supposed to be a necessary preparation for the career of a statesman. Tycho, however, speedily made it plain to his teachers that though he was an ardent student, yet the things which interested him were the movements of the heavenly bodies and not the subtleties of metaphysics.

On the 21st October, 1560, an eclipse of the sun occurred, which was partially visible at Copenhagen. Tycho, boy though he was, took the utmost interest in this event. His ardour and astonishment in connection with the circumstance were chiefly excited by the fact that the time of the occurrence of the phenomenon could be predicted with so much accuracy. Urged by his desire to understand the matter thoroughly, Tycho sought to procure some book which might explain what he so greatly wanted to know. In those days books of any kind were but few and scarce, and scientific books were especially unattainable. It so happened, however, that a Latin version of Ptolemy's astronomical works had appeared a few years before the eclipse took place, and Tycho managed to buy a copy of this book, which was then the chief authority on celestial matters. Young as the boy astronomer was, he studied hard, although perhaps not always successfully, to understand Ptolemy, and to this day his copy of the great work, copiously annotated and marked by the schoolboy hand, is preserved as one of the chief treasures in the library of the University at Prague.

After Tycho had studied for about three years at the University of Copenhagen, his uncle thought it would be better to send him, as was usual in those days, to complete his education by a course of study in some foreign university. The uncle cherished the hope that in this way the attention of the young astronomer might be withdrawn from the study of the stars and directed in what appeared to him a more useful way. Indeed, to the wise heads of those days, the pursuit of natural science deemed so much waste of good time which might otherwise be devoted to logic or rhetoric or some other branch of study more in vogue at that time. To assist in this attempt to wean Tycho from his scientific tastes, his uncle chose as a tutor to accompany him, an intelligent and upright young man named Vedel, who was four years senior to his pupil, and accordingly, in 1562, we find the pair taking up their abode at the University of Leipzig.

The tutor, however, soon found that he had undertaken a most hopeless task. He could not succeed in imbuing Tycho with the slightest taste for the study of the law or the other branches of knowledge which were then thought so desirable. The stars, and nothing but the stars, engrossed the attention of his pupil. We are told that all the money he could obtain was spent secretly in buying astronomical books and instruments. He learned the name of the stars from a little globe, which he kept hidden from Vedel, and only ventured to use during the latter's absence. No little friction was at first caused by all this, but in after years a fast and enduring friendship grew up between Tycho and his tutor, each of whom learned to respect and to love the other.

Before Tycho was seventeen he had commenced the difficult task of calculating the movements of the planets and the places which, they occupied on the sky from time to time. He was not a little surprised to find that the actual positions of the planets differed very widely from those which were assigned to them by calculations from the best existing works of astronomers. With the insight of genius he saw that the only true method of investigating the movements of the heavenly bodies would be to carry on a protracted series of measurements of their places. This, which now seems to us so obvious, was then an entirely new doctrine. Tycho at once commenced regular observations in such fashion as he could. His first instrument was, indeed, a very primitive one, consisting of a simple pair of compasses, which he used in this way. He placed his eye at the hinge, and then opened the legs of the compass so that one leg pointed to one star and the other leg to the other star. The compass was then brought down to a divided circle, by which means the number of degrees in the apparent angular distance of the two stars was determined.

- ROBERT STAWELL BALL

His next advance in instrumental equipment was to provide himself with the contrivance known as the "cross-staff, "with which he used to observe the stars whenever opportunity offered. It must, of course, be remembered that in those days there were no telescopes. In the absence of optical aid, such as lenses afford the modern observers, astronomers had to rely on mechanical appliances alone to measure the places of the stars. Of such appliances, perhaps the most ingenious was one known before Tycho's time, which we have represented in the adjoining figure.

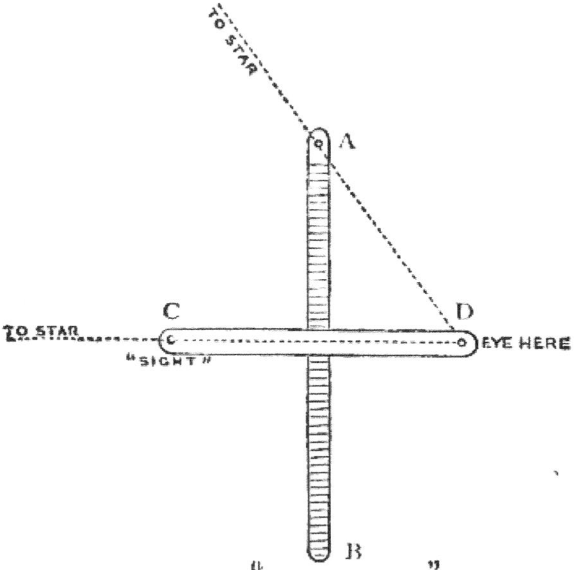

Let us suppose that it be desired to measure the angle between two stars, then if the angle be not too large it can be determined in the following manner. Let the rod A B be divided into inches and parts of an inch, and let another rod, CD, slide up and down along AB in such a way that the two al-ways remain perpendicular to each other. "Sights," like those on a rifle, are placed at A and C, and there is a pin at D. It will easily be seen that, by sliding the movable bar along the fixed one, it must always be possible when the stars are not too far apart to bring the sights into such positions that one star can be seen along D C and the other along D A. This having been accomplished, the length from A to the cross-bar is read off on the scale, and then, by means of a table previously prepared, the value of the required angular distance is obtained. If the angle between the two stars were greater than it would be possible to measure in the way already described, then there was a provision by which the pin at D might be moved along C D into some other position, so as to bring the angular distance of the stars within the range of the instrument.

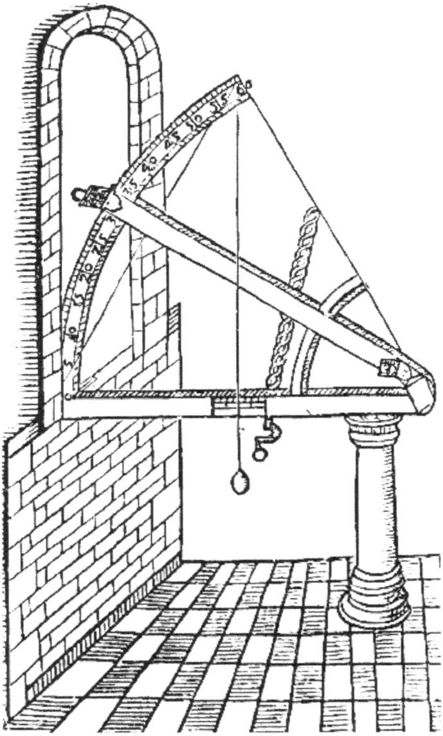

Tycho's "New Star "Sextant of 1572

(The arms, of walnut-wood, are about $5^{1/2}$ ft. long.)

No doubt the cross-staff is a very primitive contrivance, but when handled by one so skilful as Tycho it afforded results of considerable accuracy. I would recommend any reader who may have a taste for such pursuits to construct a cross-staff for himself, and see what measurements he can accomplish with its aid.

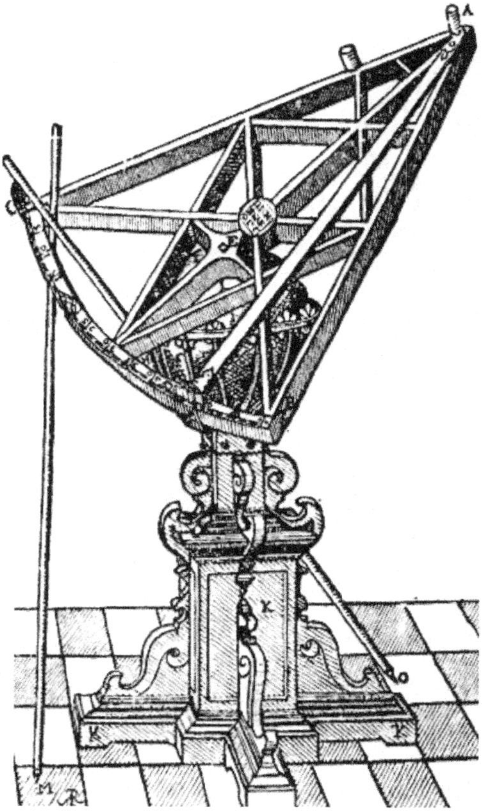

Tycho's Trigonic Sextant.

(The arms, *A B* and *A C,* are about $5^{1/2}$ ft. long.)

To employ this little instrument Tycho had to evade the vigilance of his conscientious tutor, who felt it his duty to interdict all being a frivolous waste of time. It was when Vedel was asleep that Tycho managed to escape with his cross-staff and measure the places of the heavenly bodies. Even at this early age Tycho used to conduct his observations on those thoroughly sound principles which lie at the foundation of all accurate modern astronomy. Recognising the inevitable errors of workmanship in his little instrument, he ascertained their amount and allowed for their influence on the results which he deduced. This principle, employed by the boy with his cross-staff in 1564, is employed at the present day by the Astronomer Royal at Greenwich with the most superb instruments that the skill of modern opticians has been able to construct.

After the death of his uncle, when Tycho was nineteen years of age, it appears that the young philosopher was no longer interfered with in so far as the line which his studies were to take was concerned. Always of a somewhat restless temperament, we now find that he shifted his abode to the University of Rostock, where he speedily made himself notable in connection with an eclipse of the moon on 28th October, 1566. Like every other astronomer of those days, Tycho had always associated astronomy with astrology. He considered that the phenomena of the heavenly bodies always had some significance in connection with human affairs. Tycho was also a poet, and in the united capacity of poet, astrologer, and astronomer, he posted up some verses in the college at Rostock announcing that the lunar eclipse was a prognostication of the death of the great Turkish Sultan, whose mighty deeds at that time filled men's minds. Presently news did arrive of the death of the Sultan, and Tycho was accordingly triumphant; but a little later it appeared that the decease had taken place *before* the eclipse, a circumstance which caused many a laugh at Tycho's expense.

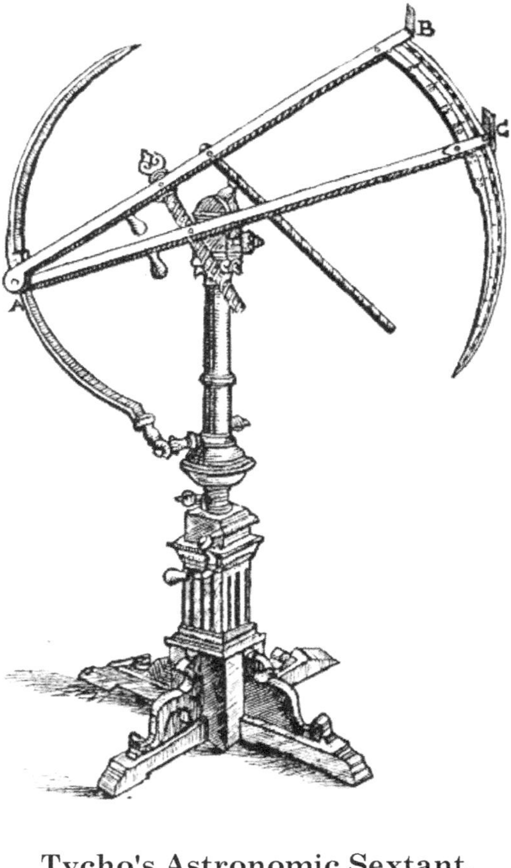

Tycho's Astronomic Sextant.
(Made of steel; the arms, A B, A C, measure 4 ft.)

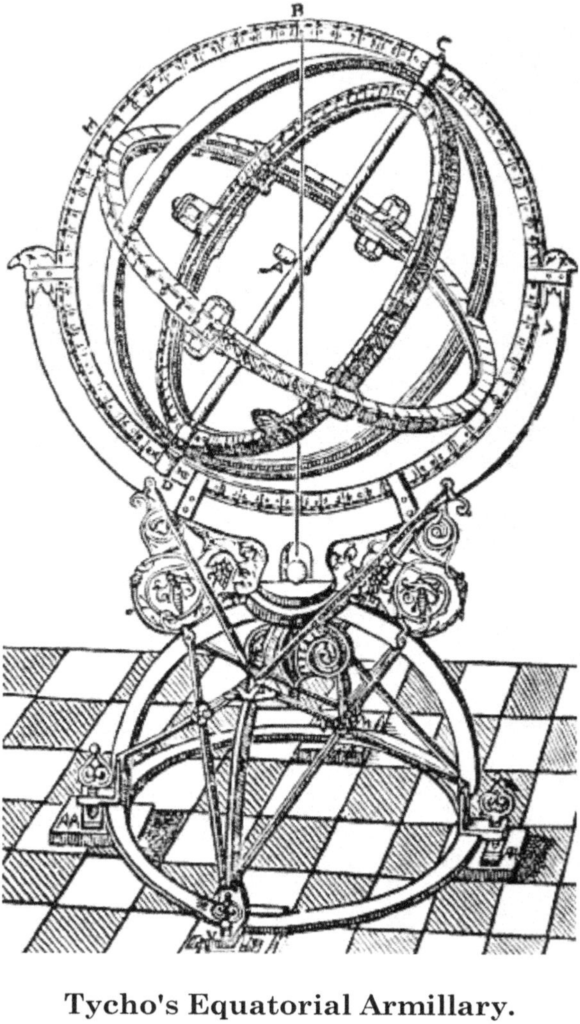

Tycho's Equatorial Armillary.

(The meridian circle, *E B C A D,* made of solid steel, is nearly 6 ft. in diameter.)

Tycho being of a somewhat turbulent disposition, it appears that, while at the University of Rostock, he had a serious quarrel with another Danish nobleman. We are not told for certain what was the cause of the dispute. It does not, however, seem to have had any more romantic origin than a difference of opinion as to which of them knew the more mathematics. They fought, as perhaps it was becoming for two astronomers to fight, under the canopy of heaven in utter darkness, at the dead of night, and the duel was honourably terminated when a slice was taken off Tycho's nose by the insinuating sword of his antagonist. For the repair of this injury the ingenuity of the great instrument-maker was here again useful, and he made a substitute for his nose "with a composition of gold and silver." The imitation was so good that it is declared to have been quite equal to the original. Dr. Lodge, however, pointedly observes that it does not appear whether this remark was made by a friend or an enemy.

The next few years Tycho spent in various places ardently pursuing somewhat varied branches of scientific study. At one time we hear of him assisting an astronomical alderman, in the ancient city of Augsburg, to erect a tremendous wooden machine—a quadrant of 19-feet radius—to be used in observing the heavens. At another time we learn that the King of Denmark had recognised the talents of his illustrious subject, and promised to confer on him a pleasant sinecure in the shape of a canonry, which would assist him with the means for indulging his scientific pursuits. Again we are told that Tycho is pursuing experiments in chemistry with the greatest energy, nor is this so incompatible as might at first be thought with his devotion to astronomy. In those early days of knowledge the different sciences seemed bound together by mysterious bonds. Alchemists and astrologers taught that the several planets were correlated in some mysterious manner with the several metals. It was, therefore, hardly surprising that Tycho should have included a study of the properties of the metals in the programme of his astronomical work.

An event, however, occurred in 1572 which stimulated Tycho's astronomical labours, and started him on his life's work. On the 11th of November in that year, he was returning home to supper after a day's work in his laboratory, when he happened to lift his face to the sky, and there he beheld a brilliant new star. It was in the constellation of Cassiopeia, and occupied a position in which there had certainly been no bright star visible when his attention had last been directed to that part of the heavens. Such a phenomenon was so startling that he found it hard to trust the evidence of his senses. He thought he must be the subject of some hallucination.

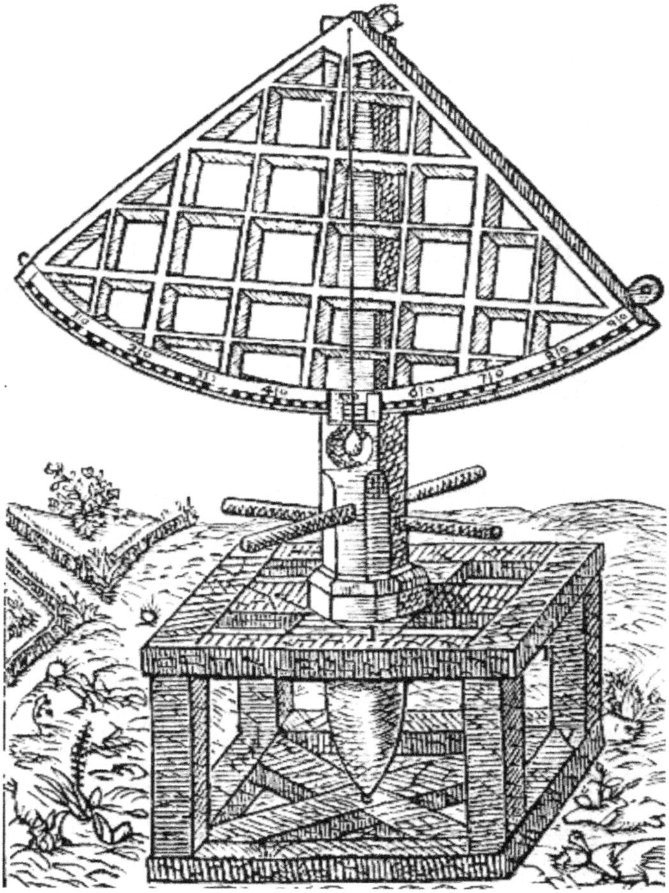

The great Augsburg Quadrant.
(Built of heart of oak; the radii about 19 ft.)

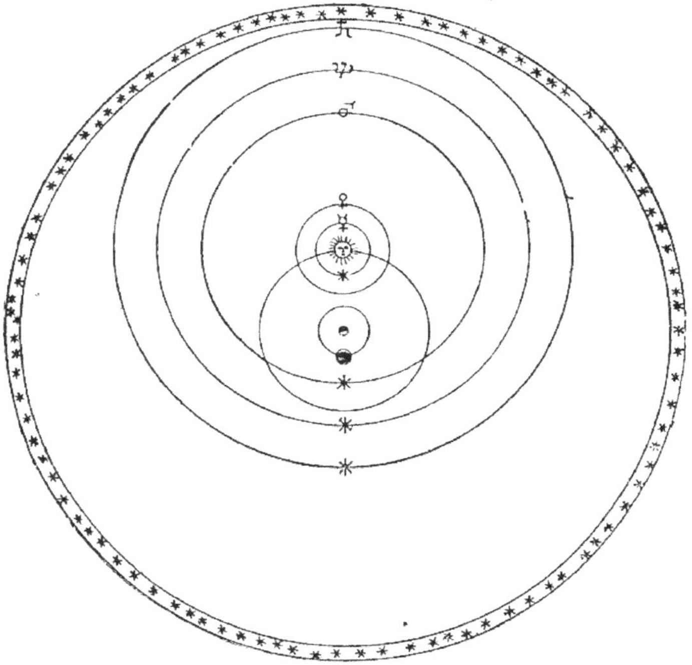

Tycho's
"New Scheme of the Terrestrial System", 1577.

He therefore called to the servants who were accompanying him, and asked them whether they, too, could see a brilliant object in the direction in which he pointed. They certainly could, and thus he became convinced that this marvellous object was no mere creation of the fancy, but a veritable celestial body—a new star of surpassing splendour which had suddenly burst forth. In these days of careful scrutiny of the heavens, we are accustomed to the occasional outbreak of new stars. It is not, however, believed that any new star which has ever appeared has displayed the same phenomenal brilliance as was exhibited by the star of 1572.

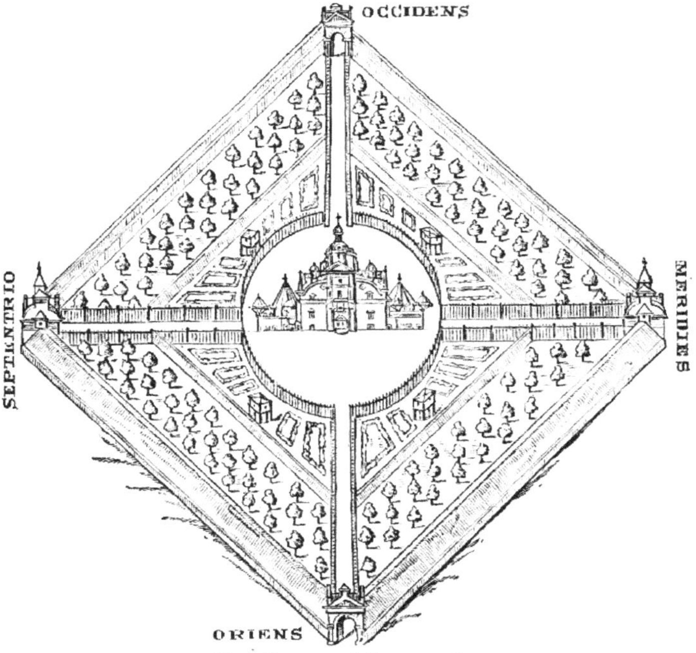

Uraniborg and its grounds.

Uraniborg and its grounds.

\- ROBERT STAWELL BALL

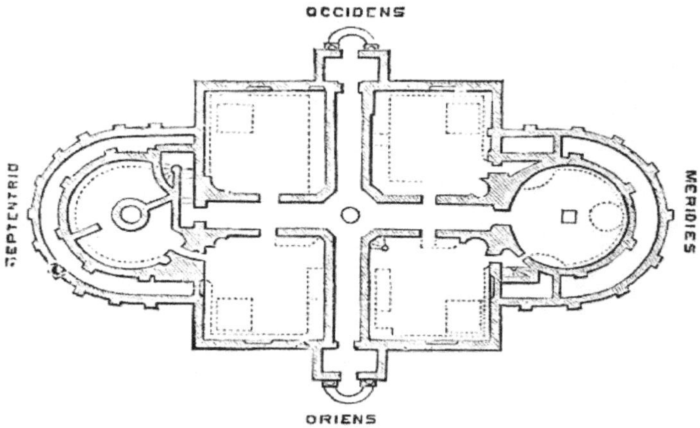

Ground-plan of the Observatory.

This object has a value in astronomy far greater than might at first appear. It is true, in one sense, that Tycho discovered the new star, but it is equally true, in a different sense, that it was the new star which discovered Tycho. Had it not been for this opportune apparition, it is quite possible that Tycho might have found a career in some direction less beneficial to science than that which he ultimately pursued.

When he reached his home on this memorable evening, Tycho immediately applied his great quadrant to the measurement of the place of the new star. His observations were specially directed to the determination of the distance of the object. He rightly conjectured that if it were very much nearer to us than the stars in its vicinity, he distance of the brilliant body might be determined in a short time by the apparent changes in its distance from the surrounding points. It was speedily demonstrated that the new star could not be as near as the moon, by the simple fact that its apparent place, as compared with the stars in its neighbourhood, was not appreciably altered when it was observed below the pole, and again above the pole at an interval of twelve hours. Such observations were possible, inasmuch as the star was bright enough to be seen in full daylight.

- ROBERT STAWELL BALL

The Observatory of Uraniborg, Island of Hven.

Tycho thus showed conclusively that the body was so remote that the diameter of the earth bore an insignificant ratio to the star's distance. His success in this respect is the more noteworthy when we find that many other observers, who studied the same object, came to the erroneous conclusion that the new star was quite as near as the moon, or even much nearer. In fact, it may be said, that with regard to this object Tycho discovered everything which could possibly have been discovered in the days before telescopes were invented. He not only proved that the star's distance was too great for measurement, but he showed that it had no proper motion on the heavens. He recorded the successive changes in its brightness from week to week, as well as the fluctuations in hue with which the alterations in lustre were accompanied.

It seems, nowadays, strange to find that such thoroughly scientific observations of the new star as those which Tycho made, possessed, even in the eyes of the great astronomer himself, a profound astrological significance. We learn from Dr. Dreyer that, in Tycho's opinion, "the star was at first like Venus and Jupiter, and its effects will therefore, first, be pleasant; but as it then became like Mars, there will next come a period of wars, seditions, captivity, and death of princes, and destruction of cities, together with dryness and fiery meteors in the air, pestilence, and venomous snakes. Lastly, the star became like Saturn, and thus will finally come a time of want, death, imprisonment, and all kinds of sad things!" Ideas of this kind were, however, universally entertained. It seemed, indeed, obvious to learned men of that period that such an apparition must forebode startling events. One of the chief theories then held was, that just as the Star of Bethlehem announced the first coming of Christ, so the second coming, and the end of the world, was heralded by the new star of 1572.

The researches of Tycho on this object were the occasion of his first appearance as an author. The publication of his book was, however, for some time delayed by the urgent remonstrances of his friends, who thought it was entirely beneath the dignity of a nobleman to condescend to write a book. Happily, Tycho determined to brave the opinion of his order; the book appeared, and was the first of a series of great astronomical productions from the same pen.

The fame of the noble Dane being now widespread, the King of Denmark entreated him to return to his native country, and to deliver a course of lectures on astronomy in the University of Copenhagen. With some reluctance he consented, and his introductory oration has been preserved. He dwells, in fervent language, upon the beauty and the interest of the celestial phenomena.

Effigy on Tycho's Tomb at Prague.

He points out the imperative necessity of continuous and systematic observation of the heavenly bodies in order to extend our knowledge, He appeals to the practical utility of the science, for what civilised nation could exist without having the means of measuring time? He sets forth how the study of these beautiful objects "exalts the mind from earthly and trivial things to heavenly ones;" and then he winds up by assuring them that "a special use of astronomy is that it enables us to draw conclusions from the movements in the celestial regions as to human fate."

An interesting event, which occurred in 1572, distracted Tycho's attention from astronomical matters. He fell in love. The young girl on whom his affections were set appears to have sprung from humble origin. Here again his august family friends sought to dissuade him from a match they thought unsuitable for a nobleman. But Tycho never gave way in anything. It is suggested that he did not seek a wife among the highborn dames of his own rank from the dread that the demands of a fashionable lady would make too great an inroad on the time that he wished to devote to science. At all events, Tycho's union seems to have been a happy one, and he had a large family of children; none of whom, however, inherited their father's talents.

Tycho had many scientific friends in Germany, among whom his work was held in high esteem.

Tycho's Mural Quadrant Picture, Uraniborg.

The treatment that he there met with seemed to him so much more encouraging than that which he received in Denmark that he formed the notion of emigrating to Basle and making it his permanent abode. A whisper of this intention was conveyed to the large-hearted King of Denmark, Frederick II. He wisely realised how great would be the fame which would accrue to his realm if he could induce Tycho to remain within Danish territory and carry on there the great work of his life. A resolution to make a splendid proposal to Tycho was immediately formed. A noble youth was forthwith despatched as a messenger, and ordered to travel day and night until he reached Tycho, whom he was to summon to the king. The astronomer was in bed on the morning of 11th February, 1576, when the message was delivered. Tycho, of course, set off at once, and had an audience of the king at Copenhagen. The astronomer explained that what he wanted was the means to pursue his studies unmolested, whereupon the king offered him the Island of Hven, in the Sound near Elsinore. There he would enjoy all the seclusion that he could desire. The king further promised that he would provide the funds necessary for building a house and for founding the greatest observatory that had ever yet been reared for the study of the heavens. After due deliberation and consultation with his friends, Tycho accepted the king's offer. He was forthwith granted a pension, and a deed was drawn up formally assigning the Island of Hven to his use all the days of his life.

The foundation of the famous castle of Uraniborg was laid on 30th August, 1576. The ceremony was a formal and imposing one, in accordance with Tycho's ideas of splendour. A party of scientific friends had assembled, and the time had been chosen so that the heavenly bodies were auspiciously placed. Libations of costly wines were poured forth, and the stone was placed with due solemnity. The picturesque character of this wonderful temple for the study of the stars may be seen in the figures with which this chapter is illustrated.

One of the most remarkable instruments that has ever been employed in studying the heavens was the mural quadrant which Tycho erected in one of the apartments of Uraniborg. By its means the altitudes of the celestial bodies could be observed with much greater accuracy than had been previously attainable. This wonderful contrivance is represented on the preceding page. It will be observed that the walls of the room are adorned by pictures with a lavishness of decoration not usually to be found in scientific establishments.

A few years later, when the fame of the observatory at Hven became more widely spread, a number of young men flocked to Tycho to study under his direction. He therefore built another observatory for their use in which the instruments were placed in subterranean rooms of which only the roofs appeared above the ground. There was a wonderful poetical inscription over the entrance to this underground observatory, expressing the astonishment of Urania at finding, even in the interior of the earth, a cavern devoted to the study of the heavens. Tycho was indeed always fond of versifying, and he lost no opportunity of indulging this taste whenever an occasion presented itself.

Around the walls of the subterranean observatory were the pictures of eight astronomers, each with a suitable inscription—one of these of course represented Tycho himself, and beneath were written words to the effect that posterity should judge of his work. The eighth picture depicted an astronomer who has not yet come into existence. Tychonides was his name, and the inscription expresses the modest hope that when he does appear he will be worthy of his great predecessor. The vast expenses incurred in the erection and the maintenance of this strange establishment were defrayed by a succession of grants from the royal purse.

For twenty years Tycho laboured hard at Uraniborg in the pursuit of science. His work mainly consisted in the determination of the places of the moon, the planets, and the stars on the celestial sphere. The extraordinary pains taken by Tycho to have his observations as accurate as his instruments would permit, have justly entitled him to the admiration of all succeeding astronomers. His island home provided the means of recreation as well as a place for work. He was surrounded by his family, troops of friends were not wanting, and a pet dwarf seems to have been an inmate of his curious residence. By way of change from his astronomical labours he used frequently to work with his students in his chemical laboratory. It is not in-deed known what particular problems in chemistry occupied his attention. We are told, however, that he engaged largely in the production of medicines, and as these appear to have been dispensed gratuitously there was no lack of patients.

Tycho's imperious and grasping character frequently brought him into difficulties, which seem to have increased with his advancing years. He had ill-treated one of his tenants on Hven, and an adverse decision by the courts seems to have greatly exasperated the astronomer. Serious changes also took place in his relations to the court at Copenhagen. When the young king was crowned in 1596, he reversed the policy of his predecessor with reference to Hven. The liberal allowances to Tycho were one after another withdrawn, and finally even his pension was stopped. Tycho accordingly abandoned Hven in a tumult of rage and mortification. A few years later we find him in Bohemia a prematurely aged man, and he died on the 24th October, 1601.

- ROBERT STAWELL BALL

COMPLIMENTARY MATERIAL

A BRIEF BIOGRAPHICAL SKETCH

TYCHO BRAHE

- ROBERT STAWELL BALL

Tycho Brahe

(December 14, 1546 – October 24, 1601)

COMPLIMENTARY MATERIAL

A BRIEF BIOGRAPHICAL SKETCH

TYCHO BRAHE

Tycho Brahe (December 14, 1546 – October 24, 1601), was a Danish nobleman known for his accurate and comprehensive astronomical and planetary observations. He was born in the then Danish peninsula of Scania. Well known in his lifetime as an astronomer, astrologer and alchemist, he has been described as "the first competent mind in modern astronomy to feel ardently the passion for exact empirical facts." His observations were some five times more accurate than the best available observations at the time.

An heir to several of Denmark's principal noble families, he received a comprehensive education. He took an interest in astronomy and in the creation of more accurate instruments of measurement. As an astronomer, Tycho worked to combine what he saw as the geometrical benefits of the Copernican system with the philosophical benefits of the Ptolemaic system into his own model of the universe, the Tychonic system. His system correctly saw the Moon as orbiting Earth, and the planets as orbiting the Sun, but erroneously considered the Sun to be orbiting the Earth. Furthermore, he was the last of the major naked eye astronomers, working without telescopes for his observations. In his De nova stella (On the New Star) of 1573, he refuted the Aristotelian belief in an unchanging celestial realm. His precise measurements indicated that "new stars" (stellae novae, now known as supernovae), in particular that of 1572, lacked the parallax expected in sublunar phenomena and were therefore not tailless comets in the atmosphere as previously believed but were above the atmosphere and beyond the moon. Using similar measurements he showed that comets were also not atmospheric phenomena, as previously thought, and must pass through the supposedly immutable celestial spheres.

King Frederick II granted Tycho an estate on the island of Hven and the funding to build Uraniborg, an early research institute, where he built large astronomical instruments and took many careful measurements, and later Stjerneborg, underground, when he discovered that his instruments in Uraniborg were not sufficiently steady. On the island (where he behaved autocratically toward the residents) he founded manufactories, such as a paper mill, to provide material for printing his results. After disagreements with the new Danish king, Christian IV, in 1597, he went into exile, and was invited by the Bohemian king and Holy Roman Emperor Rudolph II to Prague, where he became the official imperial astronomer. He built an observatory at Benátky nad Jizerou. There, from 1600 until his death in 1601, he was assisted by Johannes Kepler, who later used Tycho's astronomical data to develop his three laws of planetary motion.

Tycho's body has been exhumed twice, in 1901 and 2010, to examine the circumstances of his death and to identify the material from which his artificial nose was made. The conclusion was that his death was likely caused by a burst bladder, and not by poisoning as had been suggested, and that the artificial nose was more likely made of brass than silver or gold, as some had believed in his time.

Life

Tycho was born as heir to several of Denmark's most influential noble families and in addition to his immediate ancestry with the Brahe and the Bille families, he also counted the Rud, Trolle, Ulfstand, and Rosenkrantz families among his ancestors. Both of his grandfathers and all of his great grandfathers had served as members of the Danish king's Privy Council. His paternal grandfather and namesake Thyge Brahe was the lord of Tosterup Castle in Scania and died in battle during the 1523 Siege of Malmö during the Lutheran Reformation Wars. His maternal grandfather Claus Bille, lord to Bohus Castle and a second cousin of Swedish king Gustav Vasa, participated in the Stockholm Bloodbath on the side of the Danish king against the Swedish nobles. Tycho's father Otte Brahe, like his father a royal Privy Councilor, married Beate Bille, who was herself a powerful figure at the Danish court holding several royal land titles. Both parents are buried under the floor of Kågeröd Church, four kilometres east of Knutstorp.

1586 portrait of Tycho Brahe framed by the family shields of his noble ancestors, by Jacques de Gheyn.

Early Years

Tycho was born at his family's ancestral seat of Knutstorp Castle (Danish: Knudstrup borg; Swedish: Knutstorps borg), about eight kilometres north of Svalöv in then Danish Scania. He was the oldest of 12 siblngs, 8 of whom lived to adulthood. His twin brother died before being baptized. Tycho later wrote an ode in Latin to his dead twin, which was printed in 1572 as his first published work. An epitaph, originally from Knutstorp, but now on a plaque near the church door, shows the whole family, including Tycho as a boy.

When he was only two years old Tycho was taken away to be raised by his uncle Jørgen Thygesen Brahe and his wife Inger Oxe (sister to Peder Oxe, Steward of the Realm) who were childless. It is unclear why the Otte Brahe reached this arrangement with his brother, but Tycho was the only one of his siblings not to be raised by his mother at Knutstorp.

Instead, Tycho was raised at Jørgen Brahe's estate at Tosterup and at Tranekær on the island of Langeland, and later at Næsbyhoved Castle near Odense, and later again at the Castle of Nykøbing on the island of Falster. Tycho later wrote that Jørgen Brahe "raised me and generously provided for me during his life until my eighteenth year; he always treated me as his own son and made me his heir".

From ages 6 to 12, Tycho attended Latin school, probably in Nykøbing. At age 12, on 19 April 1559, Tycho began studies at the University of Copenhagen. There, following his uncle's wishes, he studied law, but also studied a variety of other subjects and became interested in astronomy. At the University, Aristotle was a staple of scientific theory, and Tycho likely received a thorough training in Aristotelian physics and cosmology. He experienced the solar eclipse of 21 August 1560, and was greatly impressed by the fact that it had been predicted, although the prediction based on current observational data was a day off. He realized that more accurate observations would be the key to making more exact predictions. He purchased an ephemeris and books on astronomy, including Johannes de Sacrobosco's De sphaera mundi, Petrus Apianus's Cosmographia seu descriptio totius orbis and Regiomontanus's De triangulis omnimodis.

Jørgen Thygesen Brahe, however, wanted Tycho to educate himself in order to become a civil servant, and sent him on a study tour of Europe in early 1562. 15-year old Tycho was given as mentor the 19-year-old Anders Sørensen Vedel, whom he eventually talked into allowing the pursuit of astronomy during the tour. Vedel and his pupil left Copenhagen in February 1562. On March 24 they arrived in Leipzig, where they matriculated at the Lutheran Leipzig University. In 1563 he observed a conjunction of Jupiter and Saturn, and noticed that the Copernican and Ptolemaian tables used to predict the conjunction were inaccurate. This led him to realize that progress in astronomy required systematic, rigorous observation, night after night, using the most accurate instruments obtainable. He began maintaining detailed journals of all his astronomical observations. In this period he combined the study of astronomy with astrology, laying down horoscopes for different famous personalities.

When Tycho and Vedel returned from Leipzig in 1565 Denmark was at war with Sweden, and as vice-admiral of the Danish fleet Jørgen Brahe had become a national hero for having participated in the sinking of the Swedish warship Mars during the First battle of Öland (1564). Shortly after Tycho's arrival in Denmark, Jørgen Brahe was defeated in the Action of 4 June 1565, and shortly afterwards died of a fever. Stories have it that he contracted pneumonia after a night of drinking with the Danish King Frederick II when the king fell into the water in a Copenhagen canal and Brahe jumped in after him. Brahe's possessions passed on to his wife Inger Oxe, who considered Tycho with special fondness.

Tycho's Nose

In 1566 Tycho Brahe left to study at the University of Rostock. Here he studied with professors of medicine at the university's famous medical school. Here he became interested in medical alchemy and botanical medicine.[n 2] On 29 December 1566 Tycho lost part of his nose in a sword duel against a fellow Danish nobleman, Manderup Parsberg (his third cousin). Tycho had quarreled with Parsberg at a wedding dance at Professor Lucas Bachmeister's house on 10 December, and again on the 27th, and the two ended up resolving whatever issue they were quarreling about with a duel. Though the two were later reconciled, the duel (in the dark) resulted in Tycho losing the bridge of his nose, and gaining a broad scar across his forehead. At the university he received the best possible care, and for the rest of his life he wore a prosthetic nose, said to be made of silver and gold, kept in place with a paste or glue. In November 2012, Danish and Czech researchers, after chemically analyzing "a small bone sample from the nose" from the body exhumed in 2010, reported that the prosthetic was "made out of brass".

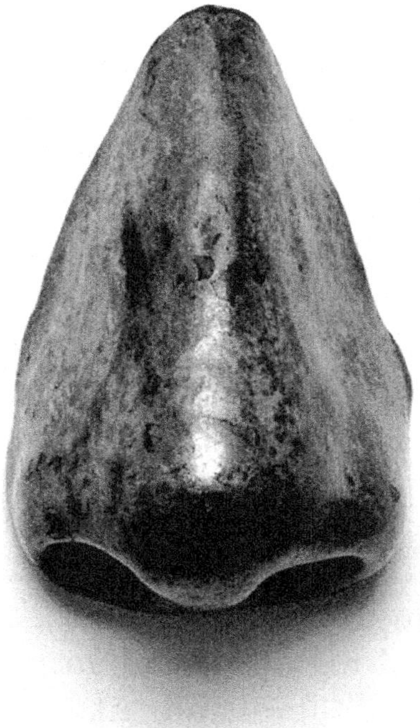

An artificial nose of the kind Tycho wore.
This particular example did not belong to Tycho.

Science and life on Uraniborg

In April 1567, Tycho returned home from his travels, with a firm intention to become a scientist. Although he had been expected to go into politics and the law, like most of his kinsmen, and although Denmark was still at war with Sweden, his family supported his decision to dedicate himself to the sciences. His father wanted him to take up law, but Tycho was allowed to travel to Rostock and then to Augsburg (where he built a great quadrant), Basel, and Freiburg. In 1568 he was appointed a canon at the Cathedral of Roskilde, a largely honorary position that would allow him to focus on his studies. At the end of 1570 he was informed of his father's ill health, so he returned to Knutstorp Castle, where his father died on 9 May 1571. The war was over, and the Danish lords soon returned to prosperity. Soon, another uncle, Steen Bille, helped him build an observatory and alchemical laboratory at Herrevad Abbey.

Marriage to Kirsten Jørgensdatter

Towards the end of 1571, Tycho fell in love with Kirsten, daughter of Jørgen Hansen, the Lutheran minister in Knudstrup. She was a commoner, and Tycho never formally married her, since if he did he would lose his noble privileges. However, Danish law permitted morganatic marriage, which meant that a nobleman and a common woman could live together openly as husband and wife for three years, and their alliance then became a legally binding marriage. Each would however maintain their social status, and any children they had together would be considered commoners, with no rights to titles, landholdings, coat of arms, or even their father's noble name. While King Frederick respected Tycho's choice of wife, himself having been unable to marry the woman he loved, many of Tycho's family members disagreed, and many churchmen would continue to hold the lack of a divinely sanctioned marriage against him. Kirsten Jørgensdatter gave birth to their first daughter, Kirstine (named after Tycho's late sister) on 12 October 1573.

Kirstine died from the plague in 1576, and Tycho wrote a heartfelt elegy for her tombstone. Together they had eight children, six of whom lived to adulthood. In 1574 they moved to Copenhagen where their daughter Magdalene was born, later the family followed him into exile. Kirsten and Tycho lived together for almost thirty years until Tycho's death.

The 1572 supernova

On 11 November 1572, Tycho observed (from Herrevad Abbey) a very bright star, now numbered SN 1572, which had unexpectedly appeared in the constellation Cassiopeia. Because it had been maintained since antiquity that the world beyond the Moon's orbit was eternally unchangeable (celestial immutability was a fundamental axiom of the Aristotelian world-view), other observers held that the phenomenon was something in the terrestrial sphere below the Moon. However, in the first instance Tycho observed that the object showed no daily parallax against the background of the fixed stars. This implied it was at least farther away than the Moon and those planets that do show such parallax. He also found the object did not change its position relative to the fixed stars over several months, as all planets did in their periodic orbital motions, even the outer planets for which no daily parallax was detectable. This suggested it was not even a planet, but a fixed star in the stellar sphere beyond all the planets. In 1573 he published a small book, De nova stella[n 3] thereby coining the term nova for a "new" star (we now classify this star as a supernova and we know that it is 7500 light-years from Earth). This discovery was decisive for his choice of astronomy as a profession. Tycho was strongly critical of those who dismissed the implications of the astronomical appearance, writing in the preface to De nova stella: "O crassa ingenia. O caecos coeli spectatores" ("Oh thick wits. Oh blind watchers of the sky"). The publication of his discovery made him a well-known name among scientists across Europe.

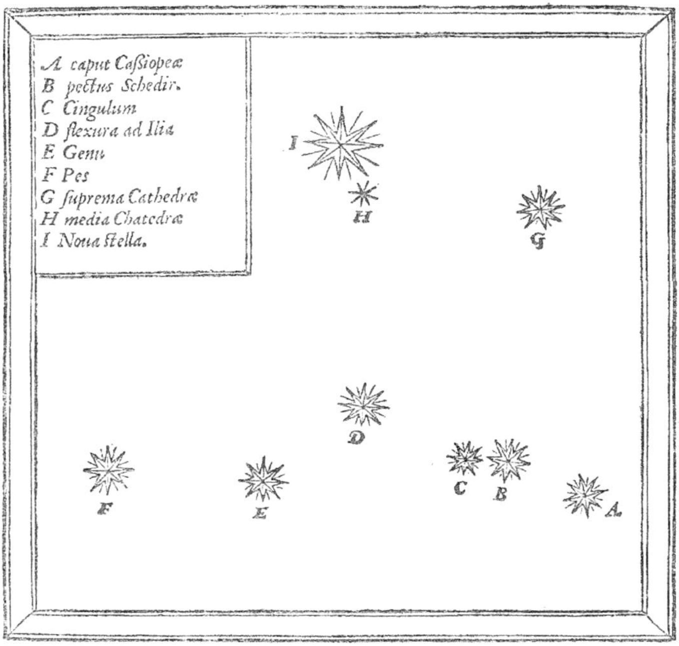

Diſtantiam verò huius ſtellæ à fixis aliquibus in hac Caßiopeiæ conſtellatione, exquiſito inſtrumento, & omnium minutorum capacj, aliquoties obſeruaui. Inueni autem eam diſtare ab ea, quæ eſt in pectore, Schedir appellata B, 7. partibus & 55. minutis : à ſuperiori verò

Star map of the constellation Cassiopeia showing the position of the supernova of 1572 (the topmost star, labelled I); from Tycho Brahe's De nova stella

---

\- ROBERT STAWELL BALL

Lord of Hven

Tycho continued with his detailed observations, often assisted by his first assistant and student, his younger sister Sophie Brahe. In 1574, Tycho published the observations made in 1572 from his first observatory at Herrevad Abbey. He then started lecturing on astronomy, but gave it up and left Denmark in spring 1575 to tour abroad. He first visited William IV, Landgrave of Hesse-Kassel's observatory at Kassel, then went on to Frankfurt, Basel and Venice, where he acted as an agent for the Danish king, contacting artisans and craftsmen whom the king wanted to work on his new palace at Elsinore. Upon his return the King wished to repay Tycho's service by offering him a position worthy of his family; he offered him a choice of lordships of militarily and economically important estates, such as the castles of Hammershus or Helsingborg. But Tycho was reluctant to take up a position as a lord of the realm, preferring to focus on his science. He wrote to his friend Johannes Pratensis, "I did not want to take possession of any of the castles our benevolent king so graciously offered me. I am displeased with society here, customary forms and the whole rubbish". Tycho secretly began to plan to move to Basel, wishing to participate in the burgeoning academic and scientific life there. But the King heard of Tycho's plans, and desiring to keep the distinguished scientist, he offered Tycho the island of Hven in Øresund and funding to set up an observatory.

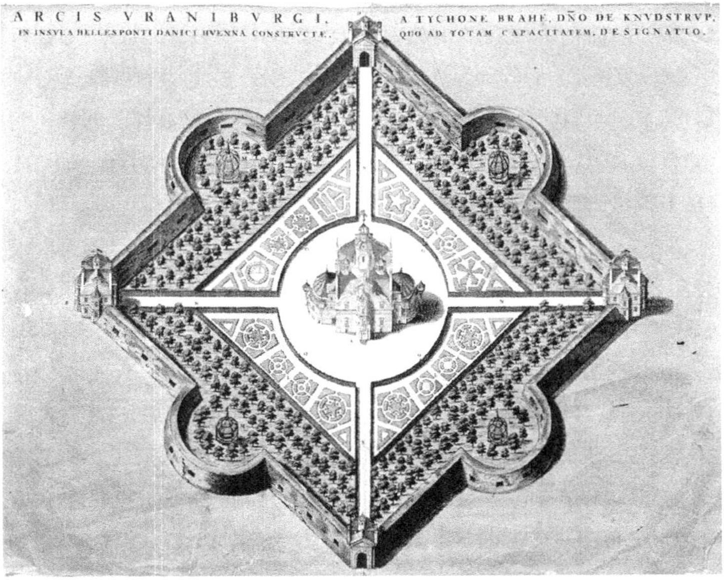

Watercolor plan of Uraniborg

Until then, Hven had been property directly under the Crown, and the 50 families on the island considered themselves to be freeholding farmers, but with Tycho Brahe's appointment as Feudal Lord of Hven this changed. Tycho took control of agricultural planning, requiring the peasants to cultivate twice as much as they had done before, and he also exacted corvée labor from the peasants for the construction of his new castle. The peasants complained about Brahe's excessive taxation and took him to court. The court established Tycho's right to levy taxes and labor, and the result was a contract detailing the mutual obligations of lord and peasants on the island.

Brahe envisioned his castle Uraniborg as a temple dedicated to the muses of arts and sciences, rather than as a military fortress; indeed it was named after Urania, the muse of astronomy. Construction began in 1576 (with a laboratory for his alchemical experiments in the cellar). Uraniborg was inspired by the Venetian architect Andrea Palladio, and was one of the first buildings in northern Europe to show influence from

Italian renaissance architecture. When he realized that the towers of Uraniborg were not adequate as observatories because of the instruments' exposure to the elements and the movement of the building, he then constructed a second underground observatory at nearby Stjerneborg in 1581. The basement included an alchemical laboratory with 16 furnaces for conducting distillations and other chemical experiments. Unusually for the time, Tycho established Uraniborg as a research centre, where almost 100 students and artisans worked from 1576 to 1597. Uraniborg also contained a printing press and a paper mill, both among the first in Scandinavia, enabling Tycho to publish his own manuscripts, on locally made paper with his own watermark. He created a system of ponds and canals to run the wheels of the paper mill. Over the years he worked on Uraniborg, Tycho was assisted by a number of students and protegés, many of whom went on to their own careers in astronomy: among them were Christian Sørensen Longomontanus, later one of the main proponents of the Tychonic model and Tycho's replacement as royal

Danish astronomer; Peder Flemløse; Elias Olsen Morsing; and Cort Aslakssøn. Tycho's instrument-maker Hans Crol also formed part of the scientific community on the island.

Tycho Brahe's large mural quadrant at Uraniborg

He observed the great comet that was visible in the Northern sky from November 1577 to January 1578. Within Lutheranism it was commonly believed that celestial objects like comets were powerful portents, announcing the coming apocalypse, and in addition to Tycho's observations several Danish amateur astronomers observed the object and published prophesies of impending doom. He was able to determine that the comet's distance to Earth was much greater than the distance of the Moon, so that the comet could not have originated in the "earthly sphere", confirming his prior anti-Aristotelian conclusions about the fixed nature of the sky beyond the Moon. He also realized that the comet's tail was always pointing away from the Sun. He calculated its diameter, mass, and the length of its tail, and speculated about the material it was made of. At this point he had not yet broken with Copernican theory, and observing the comet inspired him to try to develop an alternative Copernican model in which the Earth was immobile. The second half of his manuscript about the comet dealt with the astrological and apocalyptic aspects of the comet, and he rejected the prophesies of his competitors, instead making his own predictions of dire political events in the near future. Among his predictions was bloodshed in Moscow and the imminent fall of Ivan the Terrible by 1583.

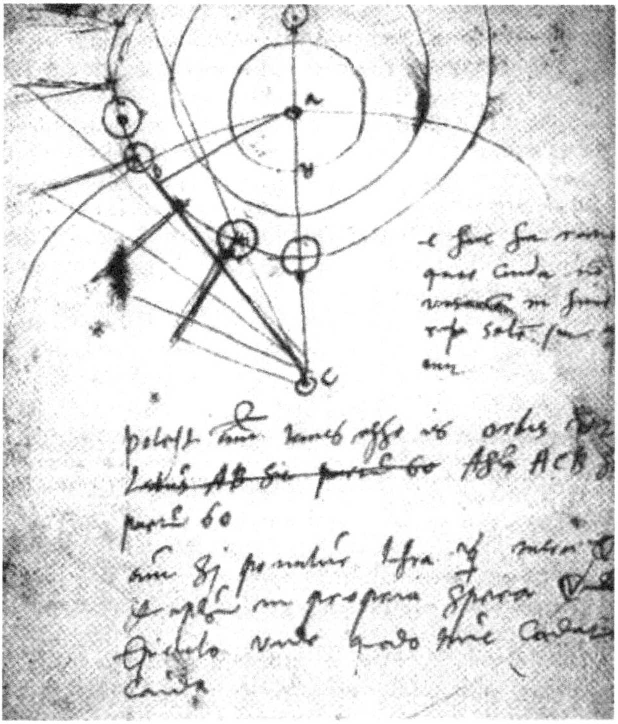

Brahe's notebook with his observations of the 1577 comet.

The support that Tycho received from the Crown was substantial, amounting to 1% of the annual total revenue at one point in the 1580s. Tycho often held large social gatherings in his castle. Pierre Gassendi wrote that Tycho also had a tame elk (moose) and that his mentor the Landgrave Wilhelm of Hesse-Kassel (Hesse-Cassel) asked whether there was an animal faster than a deer. Tycho replied that there was none, but he could send his tame elk. When Wilhelm replied he would accept one in exchange for a horse, Tycho replied with the sad news that the elk had just died on a visit to entertain a nobleman at Landskrona. Apparently during dinner the elk had drunk a lot of beer, fallen down the stairs, and died.[n 5] Among the many noble visitors to Hven was James VI of Scotland who married the Danish princess Anne. After his visit to Hven in 1590 he wrote a poem comparing Tycho Brahe with Apollon and Phaethon.

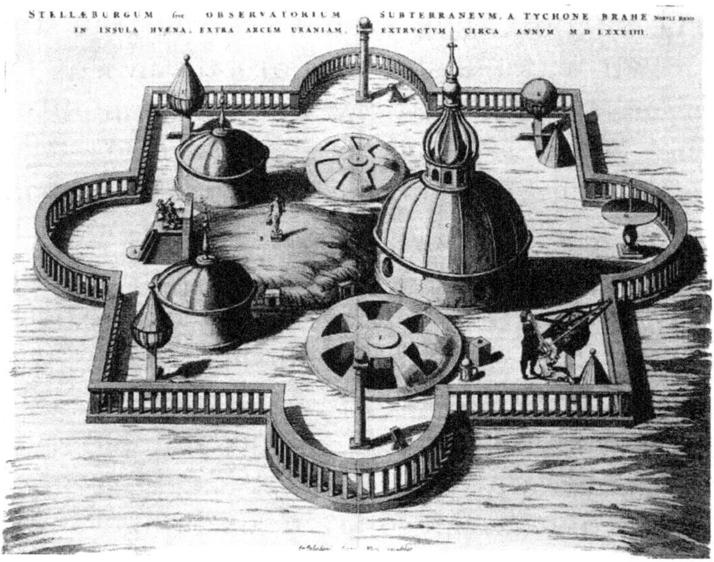

Drawing of the above ground parts of
Tycho Brahe's underground observatory
"Stjerneborg".

As part of Tycho's duties to the Crown in exchange for his estate, he fulfilled the functions of a royal astrologer. At the beginning of each year he had to present an Almanac to the court, predicting the influence of the stars on the political and economic prospects of the year. And at the birth of each prince, he prepared their horoscopes, predicting their fates. He also worked as a cartographer with his former tutor Anders Sørensen Vedel on mapping out all of the Danish realm. An ally of the king and friendly with Queen Sophie (both his mother Beate Bille and adoptive mother Inger Oxe had been her court maids), he secured a promise from the King that ownership of Hven and Uraniborg would pass to his heirs.

Publications, Correspondence and Scientific disputes

In 1588, Tycho's royal benefactor died, and a volume of Tycho's great two-volume work Astronomiae Instauratae Progymnasmata (Introduction to the New Astronomy) was published. The first volume, devoted to the new star of 1572, was not ready, because the reduction of the observations of 1572–3 involved much research to correct the stars' positions for refraction, precession, the motion of the Sun etc., and was not completed in Tycho's lifetime (it was published in Prague in 1602/03), but the second volume, titled De Mundi Aetherei Recentioribus Phaenomenis Liber Secundus (Second Book About Recent Phenomena in the Celestial World) and devoted to the comet of 1577, was printed at Uraniborg and some copies were issued in 1588. Besides the comet observations it included an account of Tycho's system of the world. The third volume was intended to treat the comets of 1580 and following years in a similar manner, but it was never published, nor even written, though a great deal of material about the comet of 1585 was put together and first published in 1845 with the observations of this comet.

While at Uraniborg, Tycho Brahe maintained correspondence with scientists and astronomers across Europe. He inquired about other astronomers' observations and shared his own technological advances to help them achieve more accurate observations. Thus his correspondence was crucial to his research. Often correspondence was not just private communication between scholars, but also a way to disseminate results and arguments and to build progress and scientific consensus. Through correspondence, Tycho Brahe was involved in several personal disputes with critics of his theories. Prominent among them were John Craig, a Scottish physician who was a strong believer in the authority of the Aristotelian worldview, and Nicolaus Reimers Baer, known as Ursus, an astronomer at the Imperial court in Prague, whom Tycho accused of having plagiarized his cosmological model. Craig refused to accept Brahe's conclusion that the comet of 1577 had to be located within the aetherial sphere rather than within the atmosphere of Earth. Craig tried to contradict Brahe by using his own observations of the comet, and by questioning his methodology. Brahe published an apologia (a defense) of his conclusions, in which he provided additional arguments, as well as condemning Craig's ideas in strong language for being incompetent. Another dispute concerned the mathematician Paul Wittich, who, after staying on Hven in 1580, taught Count Wilhelm of Kassel and his astronomer Christoph Rothmann to build copies of Brahe's instruments without permission from Brahe. In turn, Craig, who had studied with Wittich, accused Brahe of minimizing Wittich's role in developing some of the trigonometric methods used by Brahe. In his dealings with these disputes, Tycho Brahe made sure to leverage his support in the scientific community, by publishing and disseminating his own answers and arguments.

TYCHONIS
BRAHE DANI,
ASTRONOMIÆ INSTAU-
RATÆ PROGYMNASMATA.
Quorum hæc
PRIMA PARS
DE RESTITVTIONE MOTVVM
Solis & Lunæ, Stellarumque inerrantium
tractat.
ET PRÆTEREA DE ADMIRANDA
noua Stella Anno 1572. exorta luculenter agit.

ANNO  *M. DCX.*

Excudi primum cœpta Vraniburgi Daniæ,
ast Pragæ Bohemiæ absoluta.
PROSTANT
Francofurti apud Godefridum Tampachium.
Cum Cæsaris & Regum quorundam priuilegiis.

Frontispiece to the 1610 edition of Astronomiae Instauratae Progymnasmata

---

\- ROBERT STAWELL BALL

Exile and Later Years

" Denmark what is my offense? How
have I offended you my fatherland?
You may think that what I have done is wrong
But was I wrong to spread your fame abroad?
Tell me, who has done such things before?
And sung your honor to the very stars? "

- Excerpt of Tycho Brahe's Elegy to Dania

When Frederick died in 1588 his son and heir Christian IV was only 11 years old. A regency council was appointed to rule for the young prince-elect until his coronation in 1596. The head of the council (Steward of the Realm) was Christoffer Valkendorff, who disliked Tycho Brahe after a conflict between them, and hence Tycho's influence at the Danish court steadily declined. Feeling that his legacy on Hven was in peril he approached the Dowager Queen Sophie and asked her to affirm in writing her late husband's promise to endow Hven to Tycho's heirs. Nonetheless, he realized that the young king was more interested in war than in science, and was of no mind to keep his father's promise. King Christian IV followed a policy of curbing the power of the nobility by confiscating their estates to minimize their income bases, by accusing nobles of misusing their offices and of heresies against the Lutheran church. Tycho, who was known to sympathize with the Philippists (followers of Philip Melanchthon), was among the nobles who fell out of grace with the new king. The king's unfavorable disposition towards Tycho was likely also a result of efforts by several of his enemies at court to turn the king against him. Tycho's enemies included, in addition to Valkendorff, the king's doctor Peter Severinus, who also had a personal gripes with Brahe, and several gnesio-Lutheran Bishops who suspected Brahe of heresy — a suspicion motivated by his known Philippist sympathies, his pursuits in medicine and alchemy (both of which he practiced without the church's approval) and his prohibiting the local priest on Hven to include the exorcism in the baptismal ritual. Among the accusations raised against Tycho Brahe were his failure to adequately maintain the royal chapel at Roskilde, and his harshness and exploitation of the Hven peasantry.

The straw that broke the camel's back for Tycho was when a mob of commoners, possibly incited by his enemies at court, rioted in front of his house in Copenhagen. Tycho Brahe left Hven in 1597, bringing some of his instruments with him to Copenhagen, and entrusting others to a caretaker on the island. Shortly before leaving he completed his star catalogue giving the positions of 1000 stars. After some unsuccessful attempts at influencing the king to let him return, he finally acquiesced to exile, but he wrote his most famous poem Elegy to Dania in which he chided Denmark for not appreciating his genius. The instruments he had used in Uraniborg and Stjerneborg were depicted and described in detail in his book Astronomiae instauratae mechanica or Instruments for the restoration of astronomy, first published in 1598. The King sent two envoys to Hven to describe the instruments left behind by Brahe.

Unversed in astronomy, the envoys reported to the king that the large mechanical contraptions such as his large quadrant and sextant were "useless and even harmful".

From 1597 to 1598, he spent a year at the castle of his friend Heinrich Rantzau in Wandesburg outside Hamburg, and then they moved for a while to Wittenberg, where they stayed in the former home of Philip Melanchthon.

In 1599 he obtained the sponsorship of Rudolf II, Holy Roman Emperor and moved to Prague, as Imperial Court Astronomer. Tycho built a new observatory in a castle in Benátky nad Jizerou, 50 km from Prague, and worked there for one year. The emperor then brought him back to Prague, where he stayed until his death. At the imperial court even Tycho's wife and children were treated like nobility, which they had never been at the Danish court.

Tycho received financial support from several nobles in addition to the emperor, including Oldrich Desiderius Pruskowsky von Pruskow, to whom he dedicated his famous Mechanica. In return for their support, Tycho's duties included preparing astrological charts and predictions for his patrons at events such as births, weather forecasting, and astrological interpretations of significant astronomical events, such as the supernova of 1572 (sometimes called Tycho's supernova) and the Great Comet of 1577.

Relationship with Kepler

In Prague, Tycho worked closely with Johannes Kepler, his assistant. Kepler was a convinced Copernican, and considered Tycho's model to be mistaken, and derived from simple "inversion" of the Sun's and Earth's positions in the Copernican model. Together the two worked on a new star catalogue based on his own accurate positions — this catalogue became the Rudolphine Tables. At the court in Prague was also the mathematician Nicolaus Reimers (Ursus), with whom Tycho had previously corresponded, and who like

Tycho had developed a geo-heliocentric planetary model which Tycho considered to have been plagiarized from his own. Kepler had previously spoken highly of Ursus, but now found himself in the problematic position of being employed by Tycho and having to defend his employer against Ursus' accusations, even though he disagreed with both of their planetary models. In 1600 he finished the tract Apologia pro Tychone contra Ursum (defense of Tycho against Ursus). Kepler had great respect for Tycho's methods and the accuracy of his observations and considered him to be the new Hipparchus who would provide the foundation for a restoration of the science of astronomy.

Illness, Death, and Investigations

Tycho Brahe's grave in Prague, new tomb stone from 1901

Tycho suddenly contracted a bladder or kidney ailment after attending a banquet in Prague, and died eleven days later, on 24 October 1601, at the age of 54. According to Kepler's first-hand account, Tycho had refused to leave the banquet to relieve himself because it would have been a breach of etiquette. After he returned home he was no longer able to urinate, except eventually in very small quantities and with excruciating pain. The night before he died he suffered from a delirium during which he was frequently heard to exclaim that he hoped he would not seem to have lived in vain. Before dying, he urged Kepler to finish the Rudolphine Tables and expressed the hope that he would do so by adopting Tycho's own planetary system, rather than that of Copernicus. It was reported that Brahe had written his own epitaph, "He lived like a sage and died like a fool." A contemporary physician attributed his death to a kidney stone, but no kidney stones were found during an autopsy performed after his body was exhumed in 1901, and the 20th-century medical assessment is that his death is more likely to have resulted from uremia.

The investigations in the 1990s have suggested that Tycho may not have died from urinary problems, but instead from mercury poisoning. It was speculated that he had been intentionally poisoned. The two main suspects were his assistant, Johannes Kepler, whose motives would be to gain access to Brahe's laboratory and chemicals, and his cousin, Erik Brahe, at the order of friend-turned-enemy Christian IV, because of rumors that Tycho had had an affair with Christian's mother.

Tycho Brahe's grave in Prague,
new tomb stone from 1901

In February 2010 the Prague city authorities approved a request by Danish scientists to exhume the remains, and in November 2010 a group of Czech and Danish scientists from Aarhus University collected bone, hair and clothing samples for analysis. The scientists, led by Dr Jens Vellev, analyzed Tycho's beard hair once again. The team reported in November 2012 that not only was there not enough mercury present to substantiate murder, but that there were no lethal levels of any poisons present. The team's conclusion was that "it is impossible that Tycho Brahe could have been murdered" and that he "most likely died of a burst bladder". The findings were confirmed by scientists from the University of Rostock who examined a sample of Brahe's beard hairs that had been taken in 1901. Although traces of mercury were found, these were present only in the outer scales. Therefore, mercury poisoning as the cause of death was ruled out while the study suggests that the accumulation of mercury may have come from the "precipitation of mercury dust from the air during [Brahe's] long-term alchemistic activities". The hair samples contain 20–100 times the natural concentration of gold until 2 months before his death.

Tycho is buried in the Church of Our Lady before Týn, in Old Town Square near the Prague Astronomical Clock.

Career: Observing the Heavens

Observational Astronomy

Tycho's view of science was driven by his passion for accurate observations, and the quest for improved instruments of measurement drove his life's work. Tycho was the last major astronomer to work without the aid of a telescope, soon to be turned skyward by Galileo and others. Given the limitations of the naked eye for making accurate observations, he devoted many of his efforts to improving the accuracy of the existing types of instrument — the sextant and the quadrant. He designed larger versions of these instruments, which allowed him to achieve much higher accuracy. Because of the accuracy of his instruments he quickly realized the influence of wind and the movement of buildings, and instead opted to mount his instruments underground directly on the bedrock.

Tycho's observations of stellar and planetary positions were noteworthy both for their accuracy and quantity. With an accuracy approaching on arcminute, his celestial positions were much more accurate than those of any predecessor or contemporary — about 5 times more accurate than the observations of the contemporary astronomer Wilhelm of Hesse. Rawlins (1993:§B2) asserts of Tycho's Star Catalog D, "In it, Tycho achieved, on a mass scale, a precision far beyond that of earlier catalogers. Cat D represents an unprecedented confluence of skills: instrumental, observational, & computational—all of which combined to enable Tycho to place most of his hundreds of recorded stars to an accuracy of ordermag 1'!"

- ROBERT STAWELL BALL

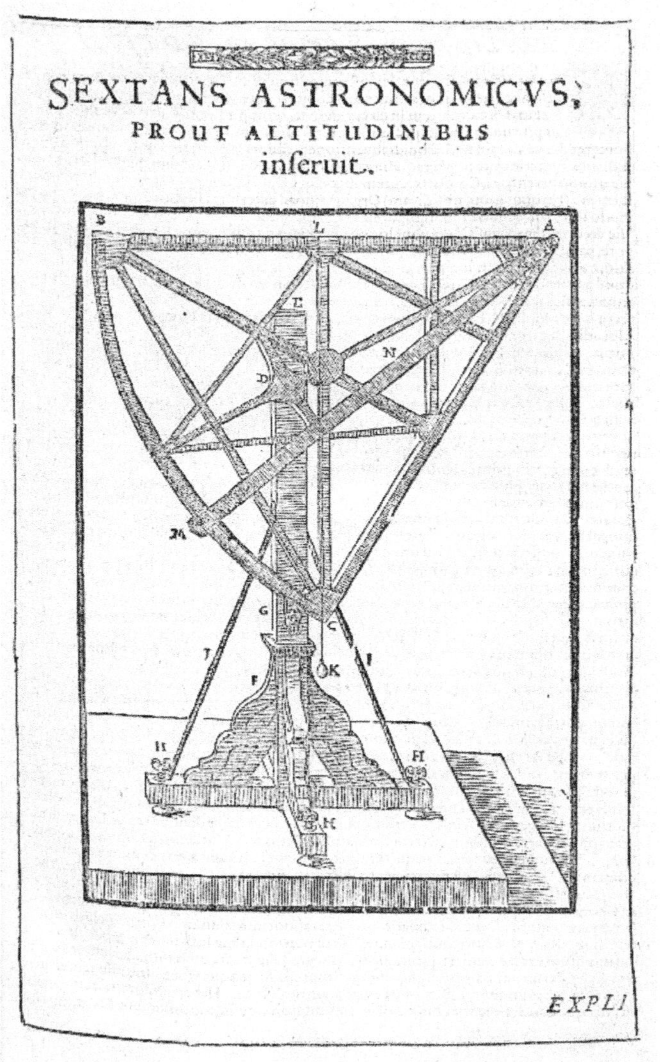

Drawing of a large sextant used by Tycho Brahe

He aspired to a level of accuracy in his estimated positions of celestial bodies of being consistently within 1 arcminute of their real celestial locations, and also claimed to have achieved this level. But in fact many of the stellar positions in his star catalogues were less accurate than that. The median errors for the stellar positions in his final published catalog were about 1'.5, indicating that only half of the entries were more accurate than that, with an overall mean error in each coordinate of around 2'. Although the stellar observations as recorded in his observational logs were more accurate, varying from 32.3" to 48.8" for different instruments, systematic errors of as much as 3' were introduced into some of the stellar positions Tycho published in his star catalog — due for instance, to his application of an erroneous ancient value of parallax and his neglect of polestar refraction. Incorrect transcription in the final published star catalogue, by scribes in Brahe's employ, was the source of even larger errors, sometimes by many degrees.

- ROBERT STAWELL BALL

Celestial objects observed near the horizon and above appear with a greater altitude than the real one, due to atmospheric refraction, and one of Tycho's most important innovations was that he worked out and published the very first tables for the systematic correction of this possible source of error. But as advanced as they were, they attributed no refraction whatever above 45 degrees altitude for solar refraction, and none for starlight above 20 degrees altitude.

To perform the huge number of multiplications needed to produce much of his astronomical data, Tycho relied heavily on the then new technique of prosthaphaeresis, an algorithm for approximating products based on trigonometric identities that predated logarithms.

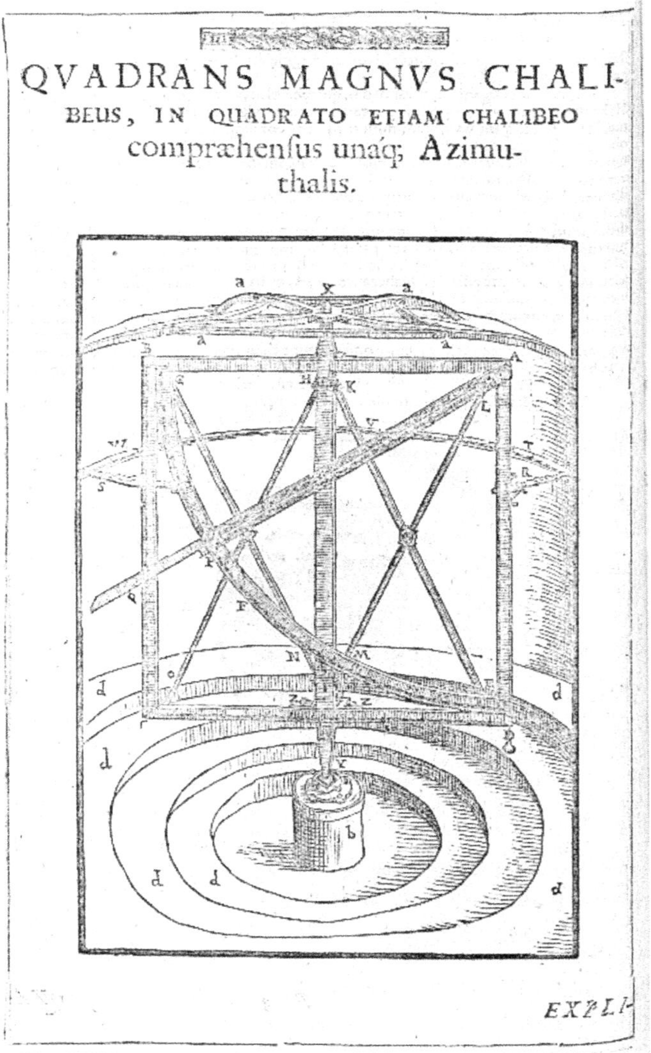

Drawing of a large quadrant used by Tycho Brahe.

- ROBERT STAWELL BALL

The Tychonic Cosmological Model

Although Tycho admired Copernicus and was the first to teach his theory in Denmark, he was unable to reconcile Copernican theory with the basic laws of Aristotelian physics, that he considered to be foundational. He was also critical of the observational data that Copernicus built his theory on, which he correctly considered to have a high margin of error. Instead Tycho Brahe proposed a "geo-heliocentric" system in which the Sun and Moon orbited the Earth, while the other planets orbited the Sun. Brahe's system had many of the same observational and computational advantages that Copernicus' system had, and both systems also could accommodate the phases of Venus, although Galilei had yet to discover them. His system provided a safe position for astronomers who were dissatisfied with older models but were reluctant to accept the heliocentrism and the Earth's motion. It gained a considerable following after 1616 when Rome decided officially that the heliocentric model was contrary to both philosophy and Scripture, and could be discussed only as a computational convenience that had no connection to fact. His system also offered a major innovation: while both the purely geocentric model and the heliocentric model as set forth by Copernicus relied on the idea of transparent rotating crystalline spheres to carry the planets in their orbits, Tycho eliminated the spheres entirely.

Kepler, as well as other Copernican astronomers, tried to persuade Tycho to adopt the heliocentric model of the solar system. Tycho advocated for a system with an immobile Earth for both scientific and religious reasons. But Tycho was not persuaded. According to Tycho, the idea of a rotating and revolving Earth would be "in violation not only of all physical truth but also of the authority of Holy Scripture, which ought to be paramount."

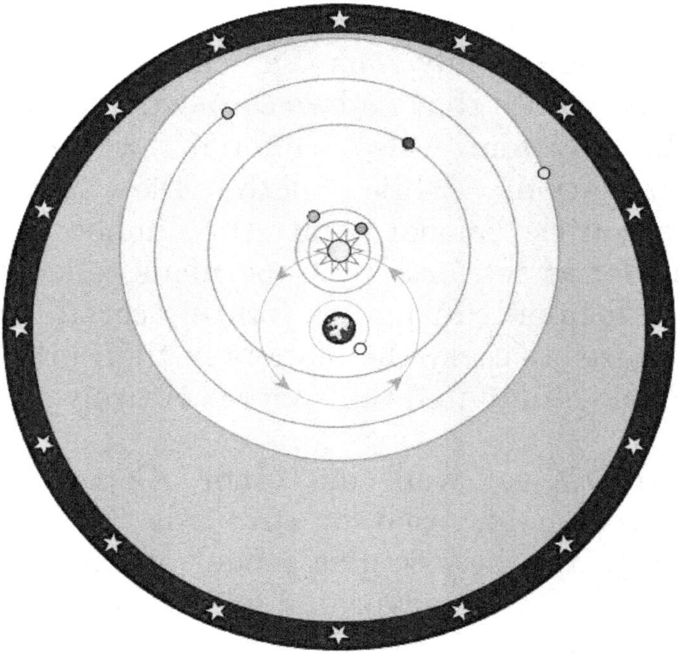

In this depiction of the Tychonic system, the objects on blue orbits (the Moon and the Sun) revolve around the Earth. The objects on orange orbits (Mercury, Venus, Mars, Jupiter, and Saturn) revolve around the Sun. Around all is a sphere of fixed stars.

With respect to physics, Tycho held that the Earth was just too sluggish and heavy to be continuously in motion. According to the accepted Aristotelian physics of the time, the heavens (whose motions and cycles were continuous and unending) were made of "Aether" or "Quintessence"; this substance, not found on Earth, was light, strong, unchanging, and its natural state was circular motion. By contrast, the Earth (where objects seem to have motion only when moved) and things on it were composed of substances that were heavy and whose natural state was rest. Accordingly, Tycho said the Earth was a "lazy" body that was not readily moved. Thus while Tycho acknowledged that the daily rising and setting of the sun and stars could be explained by the Earth's rotation, as Copernicus had said, still such a fast motion could not belong to the earth, a body very heavy and dense and opaque, but rather belongs to the sky itself whose form and subtle and constant matter are better suited to a perpetual motion, however fast.

With respect to the stars, Tycho also believed that if the Earth orbited the Sun annually there should be an observable stellar parallax over any period of six months, during which the angular orientation of a given star would change thanks to Earth's changing position. (This parallax does exist, but is so small it was not detected until 1838, when Friedrich Bessel discovered a parallax of 0.314 arcseconds of the star 61 Cygni.) The Copernican explanation for this lack of parallax was that the stars were such a great distance from Earth that Earth's orbit was almost insignificant by comparison. However, Tycho noted that this explanation introduced another problem: Stars as seen by the naked eye appear small, but of some size, with more prominent stars such as Vega appearing larger than lesser stars such as Polaris, which in turn appear larger than many others. Tycho had determined that a typical star measured approximately a minute of arc in size, with more prominent ones being two or three times as large. In writing to Christoph Rothmann, a Copernican astronomer, Tycho used basic geometry to show that, assuming a small parallax that just escaped detection, the distance to the stars in the Copernican system would have to be 700 times greater than the distance from the sun to Saturn. Moreover, the only way the stars could be so distant and still appear the sizes they do in the sky would be if even average stars were gigantic — at least as big as the orbit of the Earth, and of course vastly larger than the sun. And, Tycho said, the more prominent stars would have to be even larger still. And what if the parallax was even smaller than anyone thought, so the stars were yet more distant? Then they would all have to be even larger still. Tycho said

Deduce these things geometrically if you like, and you will see how many absurdities (not to mention others) accompany this assumption [of the motion of the earth] by inference.

Copernicans offered a religious response to Tycho's geometry: titanic, distant stars might seem unreasonable, but they were not, for the Creator could make his creations that large if He wanted. In fact, Rothmann responded to this argument of Tycho's by saying:

"What is so absurd about [an average star] having size equal to the whole [orbit of the Earth]? What of this is contrary to divine

will, or is impossible by divine Nature, or is inadmissible by infinite Nature? These things must be entirely demonstrated by you, if you will wish to infer from here anything of the absurd. These things that vulgar sorts see as absurd at first glance are not easily charged with absurdity, for in fact divine Sapience and Majesty is far greater than they understand. Grant the vastness of the Universe and the sizes of the stars to be as great as you like — these will still bear no proportion to the infinite Creator. It reckons that the greater the king, so much greater and larger the palace befitting his majesty. So how great a palace do you reckon is fitting to GOD?".

Religion played a role in Tycho's geocentrism also – he cited the authority of scripture in portraying the Earth as being at rest. He rarely used Biblical arguments alone (to him they were a secondary objection to the idea of Earth's motion) and over time he came to focus on scientific arguments, but he did take Biblical arguments seriously.

Tycho's 1587 geo-heliocentric model differed from those of other geo-heliocentric astronomers, such as Paul Wittich, Reimarus Ursus, Helisaeus Roeslin and David Origanus, in that the orbits of Mars and the Sun intersected. This was because Tycho had come to believe the distance of Mars from the Earth at opposition (that is, when Mars is on the opposite side of the sky from the Sun) was less than that of the Sun from the Earth. Tycho believed this because he came to believe Mars had a greater daily parallax than the Sun. But in 1584 in a letter to a fellow astronomer, Brucaeus, he had claimed that Mars had been further than the Sun at the opposition of 1582, because he had observed that Mars had little or no daily parallax. He said he had therefore rejected Copernicus's model because it predicted Mars would be at only two-thirds the distance of the Sun. But he apparently later changed his mind to the opinion that Mars at opposition was indeed nearer the Earth than the Sun was, but apparently without any valid observational evidence in any discernible Martian parallax. Such intersecting

Martian and solar orbits meant that there could be no solid rotating celestial spheres, because they could not possibly interpenetrate. Arguably this conclusion was independently supported by the conclusion that the comet of 1577 was superlunary, because it showed less daily parallax than the Moon and thus must pass through any celestial spheres in its transit.

Lunar Theory

Tycho's distinctive contributions to lunar theory include his discovery of the variation of the Moon's longitude. This represents the largest inequality of longitude after the equation of the center and the evection. He also discovered librations in the inclination of the plane of the lunar orbit, relative to the ecliptic (which is not a constant of about 5° as had been believed before him, but fluctuates through a range of over a quarter of a degree), and accompanying oscillations in the longitude of the lunar node. These represent perturbations in the Moon's ecliptic latitude. Tycho's lunar theory doubled the number of distinct lunar inequalities, relative to those anciently known, and reduced the discrepancies of lunar theory to about 1/5 of their previous amounts. It was published posthumously by Kepler in 1602, and Kepler's own derivative form appears in Kepler's Rudolphine Tables of 1627.

Subsequent Developments in Astronomy

Kepler used Tycho's records of the motion of Mars to deduce laws of planetary motion, enabling calculation of astronomical tables with unprecedented accuracy (the Rudolphine Tables)[n 7] and providing powerful support for a heliocentric model of the solar system.
Valentin Naboth's drawing of Martianus Capella's geo-heliocentric astronomical model (1573)

Galileo's 1610 telescopic discovery that Venus shows a full set of phases refuted the pure geocentric Ptolemaic model. After that it seems 17th century astronomy mostly converted to geo-heliocentric planetary models that could explain these phases just as well as the heliocentric model could, but without the latter's disadvantage of the failure to detect any annual stellar parallax that Tycho and others regarded as refuting it. The three main geo-heliocentric models were the Tychonic, the Capellan with just Mercury and Venus orbiting the Sun such as favoured by Francis Bacon, for example, and the extended Capellan model of Riccioli with Mars also orbiting the Sun whilst Saturn and Jupiter orbit the fixed Earth. But the Tychonic model was probably the most popular, albeit probably in what was known as 'the semi-Tychonic' version with a daily rotating Earth. This model was advocated by Tycho's ex-assistant and disciple Longomontanus in his 1622 Astronomia Danica that was the intended completion of Tycho's planetary model with his observational data, and which was regarded as the canonical statement of the complete Tychonic planetary system. Longomontanus' work was published in several editions and used by many subsequent astronomers, and through him the Tychonic system was adopted by astronomers as far away as China.

Johannes Kepler published the Rudolphine Tables containing a star catalog and planetary tables using Tycho's measurements. Hven island appears west uppermost on the base.

The ardent anti-heliocentric French astronomer Jean-Baptiste Morin devised a Tychonic planetary model with elliptical orbits published in 1650 in a simplified, Tychonic version of the Rudolphine Tables. Some acceptance of the Tychonic system persisted through the 17th century and in places until the early 18th century; it was supported (after a 1633 decree about the Copernican controversy) by "a flood of pro-Tycho literature" of Jesuit origin. Among pro-Tycho Jesuits, Ignace Pardies declared in 1691 that it was still the commonly accepted system, and Francesco Blanchinus reiterated that as late as 1728. Persistence of the Tychonic system, especially in Catholic countries, has been attributed to its satisfaction of a need (relative to Catholic doctrine) for "a safe synthesis of ancient and modern". After 1670, even many Jesuit writers only thinly disguised their Copernicanism. But in Germany, the Netherlands, and England, the Tychonic system "vanished from the literature much earlier".

James Bradley's discovery of stellar aberration, published in 1729, eventually gave direct evidence excluding the possibility of all forms of geocentrism including Tycho's. Stellar aberration could only be satisfactorily explained on the basis that the Earth is in annual orbit around the Sun, with an orbital velocity that combines with the finite speed of the light coming from an observed star or planet, to affect the apparent direction of the body observed.

Work in medicine, alchemy and astrology

Tycho Brahe also worked in medicine and alchemy. He was strongly influenced by

Paracelsus, who considered the human body to be directly influenced by celestial bodies. The paracelsian view of man as a microcosm, and astrology as the science tying together the celestial and bodily universes was also shared by Philip Melanchthon, and was precisely one of the points of contention between Melanchthon and Luther, and hence between the philippists and the gnesio-Lutherans. For Tycho Brahe there was a close connection between empiricism and natural science on one hand and religion and astrology on the other. Using his large herbal garden at Uraniborg, Tycho Brahe produced several recipes for herbal medicines, using them to treat illnesses such as fever and plague. In his own time Tycho was also famous for his contributions to medicine; his herbal medicines were in use as late as the 1900s. The expression Tycho Brahe days, in Scandinavian folklore, refers to a number of "unlucky days" that were featured in many almanacs beginning in the 1700s, but which have no direct connection to Tycho Brahe or his work. Whether because he realized that astrology was not an empirical science or because he feared religious repercussions Brahe seems to have had a somewhat ambiguous relation to his own astrological work. For example, two of his more astrological treatises one on weather predictions and an almanac were published in the names of his assistants, in spite of the fact that he worked on them personally. Some scholars have argued that he lost faith in horoscope astrology over the course of his career, and others that he simply changed his public communication on the topic as he realized that connections with astrology could influence the reception of his empirical astronomical work.

Legacy

Biographies

The first biography of Tycho Brahe, which was also the first full-length biography of any scientist, was written by Pierre Gassendi in 1654. In 1779 Tycho de Hoffmann wrote of Brahe's life in his history of the Brahe family. In 1913 Dreyer published Tycho Brahe's collected works, facilitating further research. Early modern scholarship on Tycho Brahe tended to see the shortcomings of his astronomical model, painting him as a mysticist recalcitrant in accepting the Copernican revolution, and valuing mostly his observations which allowed Kepler to formulate his laws of planetary movement. Especially in Danish scholarship Tycho Brahe was depicted as a mediocre scholar and a traitor to the nation — perhaps because of the important role in Danish historiography of Christian IV as a warrior king. In the second half of the 20th century scholars began reevaluating is significance and studies by Kristian Peder Moesgaard, Owen Gingerich, Robert Westman, Victor E. Thoren, and John R. Christianson focused on his contributions to science, and demonstrated that while he admired Copernicus he was simply unable to reconcile his basic theory of physics with the Copernican view. Christianson's work showed the influence of Tycho's Uraniborg as a training center for scientists who after studying with Brahe went on to make contributions in various scientific fields.

Monument to **Tycho Brahe** and **Kepler**
in Prague, Czech Republic.

Scientific Legacy

Although Tycho's planetary model was soon discredited, his astronomical observations were an essential contribution to the scientific revolution. The traditional view of Tycho is that he was primarily an empiricist who set new standards for precise and objective measurements. This appraisal originated in Pierre Gassendi's 1654 biography, Tychonis Brahe, equitis Dani, astronomorum coryphaei, vita. It was furthered by Johann Dreyer's biography in 1890, which was long the most influential work on Tycho. According to historian of science Helge Kragh, this assessment grew out of Gassendi's opposition to Aristotelianism and Cartesianism, and fails to account for the diversity of Tycho's activities.

Cultural Legacy

Tycho's discovery of the new star was the inspiration for Edgar Allan Poe's poem "Al Aaraaf". In 1998, Sky & Telescope magazine published an article by Donald W. Olson, Marilynn S. Olson and Russell L. Doescher arguing, in part, that Tycho's supernova was also the same "star that's westward from the pole" in Shakespeare's Hamlet.

The lunar crater Tycho is named in his honour, as is the crater Tycho Brahe on Mars and the minor planet 1677 Tycho Brahe in the asteroid belt. The bright supernova, SN 1572, is known as Tycho's Nova and the Tycho Brahe Planetarium in Copenhagen is also named after him, as is the palm genus Brahea.

Works

- *De Mundi Aetherei Recentioribus Phaenomenis Liber Secundus* (Uraniborg, 1588; Prague, 1603; Frankfurt, 1610)

- *Tychonis Brahe Astronomiae Instauratae Progymnasmata* (Prague, 1602/03; Frankfurt, 1610)

- ROBERT STAWELL BALL

ABOUT THE AUTHOR

ROBERT STAWELL BALL
(July 1, 1840 – November 25, 1913)

Sir Robert Stawell Ball was an Irish astronomer who founded the screw theory.

He was the son of naturalist Robert Ball and Amelia Gresley Hellicar. He was born in Dublin.

Ball worked for Lord Rosse from 1865 to 1867. In 1867 he became Professor of Applied Mathematics at the Royal College of Science in Dublin. There he lectured on mechanics and published an elementary account of the science.

In 1874 Ball was appointed Royal Astronomer of Ireland and Andrews Professor of Astronomy in the University of Dublin at Dunsink Observatory.

Ball contributed to the science of kinematics by delineating the screw displacement:

When Ball and the screw theorists speak of screws they no longer mean actual cylindrical objects with helical threads cut into them but the possible motion of any body whatsoever, including that of the screw independently of the nut.

Ball's treatise *The Theory of Screws* (1876) is now in the public domain. His work on screw dynamics earned him in 1879 the

Cunningham Medal of the Royal Irish Academy.

In 1882 Popular Science Monthly carried his article "*A Glimpse through the Corridors of Time*". The following year it carried his two-part article on "*The Boundaries of Astronomy*".

Ball expounded the tides in *Time and Tide*: a Romance of the Moon In 1892 he was appointed Lowndean Professor of Astronomy and Geometry at Cambridge University at the same time becoming director of the Cambridge Observatory. He was a fellow of King's College, Cambridge.

In 1900 Cambridge University Press published *A Treatise on the Theory of Screws*. That year he also published *The Story of the Heavens*. Much in the limelight, he stood as President of the Quaternion Society. He was also President of the Mathematical Association in 1900.

In 1908 he published *A Treatise on Spherical Astronomy*, which is a textbook on astronomy starting from spherical trigonometry and the celestial sphere, considering atmospheric refraction and aberration of light, and introducing basic use of a generalised instrument.

His work *The Story of the Heavens* is mentioned in the "Ithaka" chapter of Ulysses. His lectures, articles and books (e.g. *Starland* and *The Story of the Heavens*) were mostly popular and simple in style.

He died in Cambridge and was buried at the Parish of the Ascension Burial Ground in Cambridge, with his wife Lady Francis Elizabeth Ball. Their children were: Frances Amelia, Robert Steele, William Valentine (later Sir), Mary Agnetta, Charles Rowan Hamilton, and Randall Gresley (later Colonel).

LECTURES

Ball became celebrated for his popular lectures on science. He gave an estimated 2500 lectures between 1875 and 1910 in towns and cities across Britain and Ireland. In 1892, 1898 and 1900 he was invited to deliver the Royal Institution Christmas Lecture. *Astronomy; Astronomy* and *Great Chapters from the Book of Nature.*

FEEDBACK

Now that you have read the book ...

Was it interesting?

Did you enjoy what you wanted to read?
Was there any room for improvement?

Let us know at:

http://www.diamondbooks.ca/feedback

Your feedback is highly appreciated.

Thank you!

- ROBERT STAWELL BALL

Made in the USA
Las Vegas, NV
04 June 2021

24206835R00085